# WATERWORKS

# IN THE ATHENIAN AGORA

AMERICAN SCHOOL OF CLASSICAL STUDIES AT ATHENS

PRINCETON, NEW JERSEY

1968

1. Drains and Pipelines under the Hellenistic Level of the Agora. Civic Offices, Room 2, looking South.
A Cover Slabs of Great Drain, East Branch.
B Pressure Pipeline from Stone Aqueduct.
C Drain from Water Clock.
D Waste-pipe from Southwest Fountain House.

EVEN as the poet Homer (*Iliad*, XIV, 201) attributed to Ocean the birth of the Gods, so the philosopher Thales thought of water as the primary element from which all else came and on which the earth itself rested (Aristotle, *de Caelo*, 294 a 28). That such a feeling might have seemed right and proper to ancient Athenians we can now see from the excavated remnants of water-pipes and drains, so ubiquitous (1) that a pedestrian must have literally, if perhaps unconsciously, walked on water wherever he went.

The modern visitor to Greece is quite ready both to understand a myth that invokes the wrath of a god to explain the summer exhaustion of the rivers (Pausanias, II, 15, 5) and also fervently to agree with Pindar's 'Water is best' (*Olympian*, I, 1). The very real importance attached to water in ancient Athens is attested not only by words of poets and philosophers but also by the action of public men. Peisistratos (6th century B.C.) showed the enlightened quality of his despotism by adding water to the more usual 'bread and circuses'; it was he who converted the natural spring Kallirrhoe (Fair Flowing), into an elaborate fountain house (2), Enneakrounos (Nine-spouter; Thucydides, II, 15, 4–5). In the 5th century B.C. the hydraulic benefactors of Athens included two statesmen and an astronomer: Kimon made of the Academy a grove of greenery by bringing in water (Plutarch, *Kimon*, XIII, 8); Perikles generously offered to restore a springhouse (*I.G.*, I², 54); and Meton is credited with providing a fountain on Kolonos Agoraios (Scholion on Aristophanes, *Birds*, 997).

That there was at least one public official concerned with waterworks even in the earliest 5th century B.C. we learn rather casually from a story in Plutarch's Life of Themistokles (XXXI, 1). When the great general who had saved all Greece from the Persians in 480 B.C. was later exiled and visited Sardis, 'he saw in the Temple of the Mother (as part of the Persian loot) the Water Carrier, a maiden of bronze which he had dedicated when he was water commissioner in Athens, having had her made from the fines he exacted

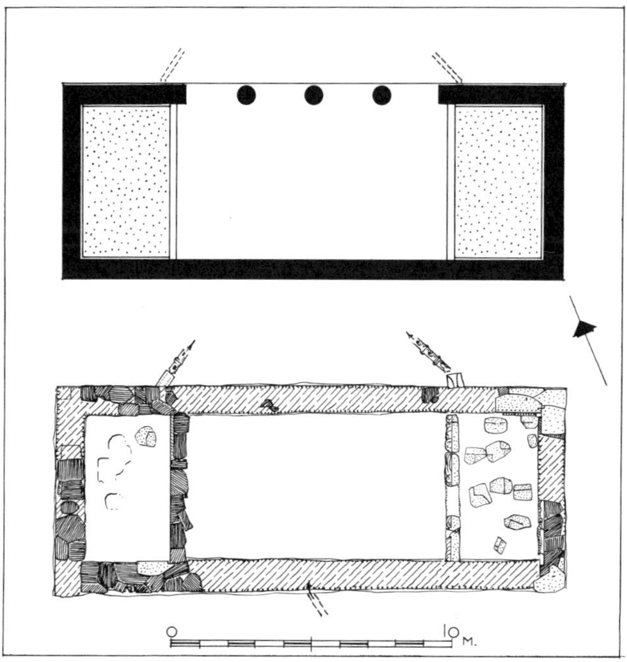


2. Enneakrounos. Late 6th Century B.C. Actual State and Restored Plans.

3. Water Carrier from the Nymphaeum. Marble. 2nd Century A.D.

from those who stole and piped off the public water' (see 3, a Roman pitcher carrier adapted from a Greek statue).

Our knowledge from the 4th century B.C. is fuller and less accidental. In his *Constitution of the Athenians* (43, 1) Aristotle reports: 'All magistrates having to do with the regular administration are chosen by lot with the exception of the Treasurer of the Military Funds, the Controllers of the Festival Fund and the Superintendent of the Fountains; these are elected by a show of hands and serve a term of four years from Panathenaia to Panathenaia.' A contemporary decree (*I.G.*, II$^2$, 338) preserved on stone echoes Aristotle and shows how hydraulic benefactions continued to be appreciated: '. . . since Pytheas, having been elected to the superintendence of the fountains, both discharged well and honorably his other duties and has now built a new fountain for the shrine of Ammon and has arranged for the bringing of water to it, it is agreed to honor Pytheas and crown him with a golden crown

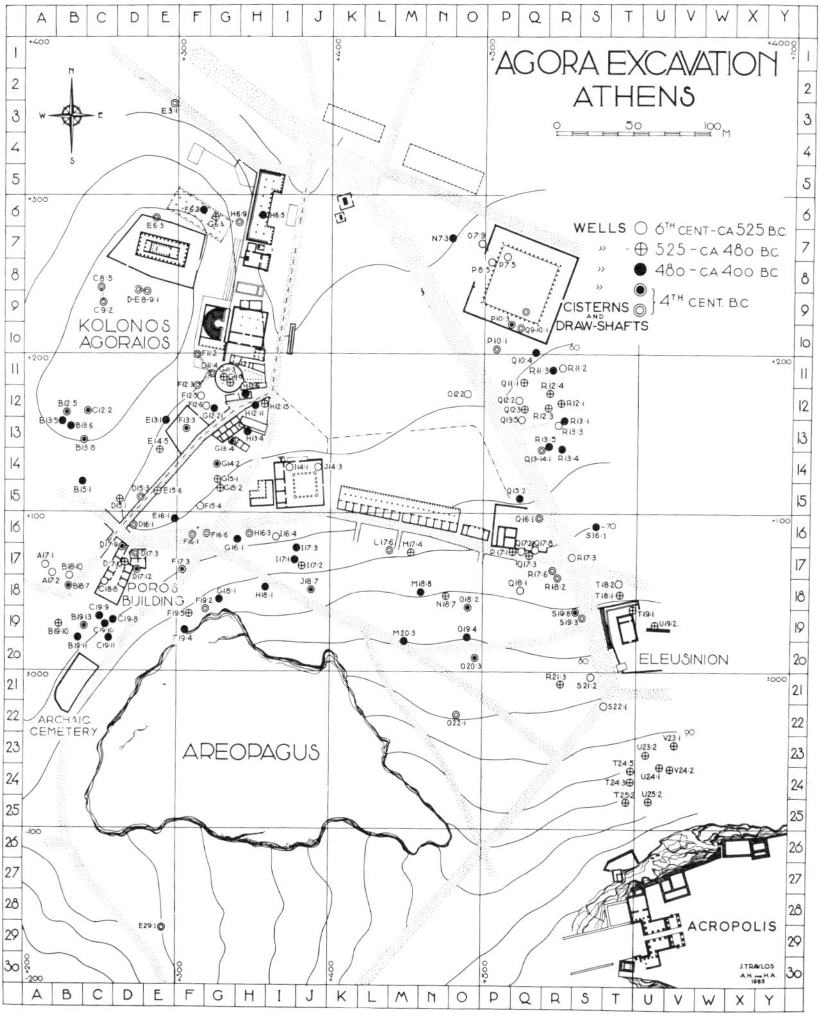


4. Three Centuries of Wells and Cisterns.

worth a thousand drachmas so that others elected to the charge of the fountains shall be ambitious on behalf of the people.'

Where did the water come from? In connection with Enneakrounos Pausanias says (I, 14, 1) that although there are wells throughout the city, this is the only spring. Whether his information is accurate for a time more than half a millennium earlier is, of course, questionable, but the Agora excavations alone have brought to light over four hundred wells (4). The plan above

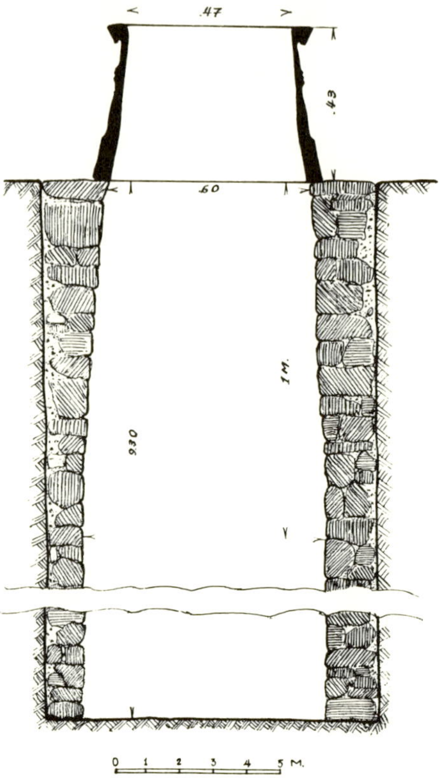


5. Terracotta Wellhead on a Stone-lined Shaft.

shows the numbers of both wells and cisterns in a mere three hundred years midway between the earliest wells of the Neolithic period and those of the 19th century A.D.

Vitruvius, the Roman engineer of the 1st century after Christ, provides (VIII, 1, 1) hints on how to locate a well: 'fall on one's face before sunrise in the general area where water is wanted and, supporting one's chin on the ground, look around the neighborhood. . . . then digging is to be carried out where moisture seems to curl upward and rise into the air.' Once dug, whether in accord with some such recipe or at random, the well was most often in the earliest periods left unlined, with toe-holes cut in the bedrock walls for both diggers and later cleaners. In the 6th century B.C. wells were occasionally lined with small-stone masonry, and, beginning in the 4th century B.C., circular drums of terracotta were often used to form a shaft of great durability. Each drum was made up

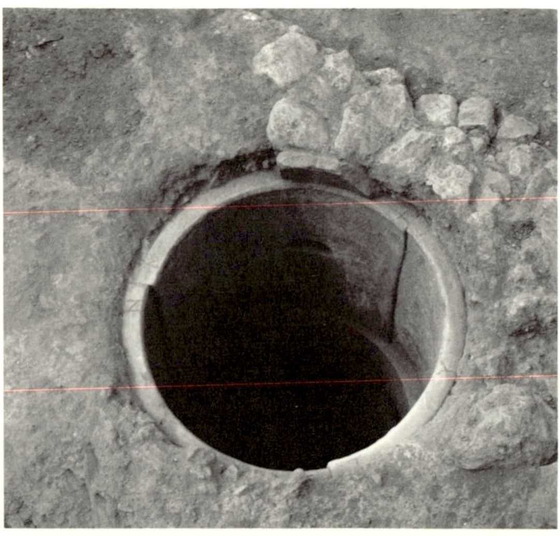

6. Tile-lined Well. Hellenistic.

of three sections, each with a hand-hold (often lunate) which must have served also as a toe-hole once the tile was in place (6).

Wells varied greatly in depth, ranging from two and one-half to thirty-seven meters and averaging about ten meters (32 feet). That this was considered no great depth in the early 6th century B.C. we learn from Plutarch's account (*Solon*, XXIII, 6) of a Solonian law: 'Since the water supply of rivers, lakes and springs was inadequate, and most people used the dug wells, he (Solon) passed a law that wherever there was a public well within half a mile this was to be used; where it was farther away, one could dig one's own: but if, having dug to a depth of 60 feet, one did not find water one was permitted to fill a five-gallon jar twice a day from one's neighbor's well; for Solon thought it right to help a man in need, but not to encourage laziness.'

Arrangements at the mouth of the well varied from time to time. Before the 6th century B.C. the neck and shoulders of a broken pithos often served. Then drums specifically designed for the purpose were made, first of terracotta (5) and later of stone. The holes which are commonly found in the

7. Excavating a Well.

8. Drawing Water from a Terracotta Wellhead. 6th Century B.C.

walls of such wellheads may have served to secure the end of the rope (8), to facilitate handling, or, above all, to receive the ends of a wooden bar from which could be suspended vessels containing food or drink requiring 're-frigeration.' The fragments of a jug found in a well with its cork still in place testify both to such a practice and to the possible unfortunate results. There is little trace of ropemarks on these wellheads, unlike those of stone, which, being higher, are usually deeply grooved by the rope drawn over the lip.

9. Wooden Sheave from a Well Pulley. 1st Century A.D.

10. Fragments of Rope from a Well. 4th Century B.C.

In the Hellenistic and Roman periods more elaborate methods were employed. A stone block may frame the mouth of the well and show cuttings for two upright members to support a horizontal beam which has slots for the axle of a pulley (11). Several wooden pulley wheels (9) have been found in wells and even some short lengths of rope (10). There are, happily, a few

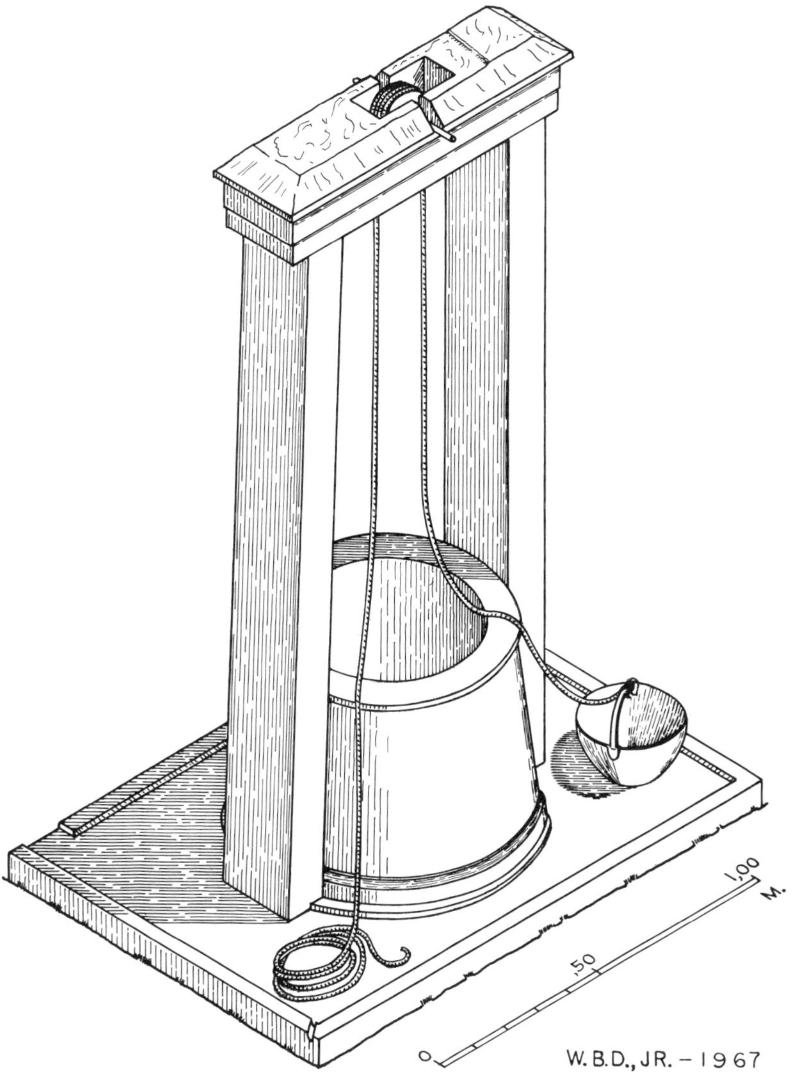


11. Wellhead and Windlass of the Hellenistic Period.

12. Wooden Bucket from a Well. 4th Century A.D.

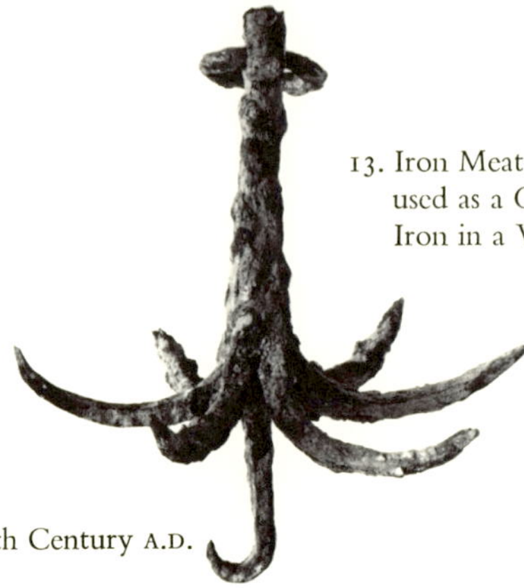

13. Iron Meat Hook used as a Grappling Iron in a Well.

iron-bound wooden buckets (12), but by far the most usual vessels employed were of earthenware, if we judge from what fell in and what so often provides the archaeologist with a stratified deposit of jugs and pitchers covering a period varying from decades to centuries. That some owners were not completely casual about vessels thus lost we learn from Aristophanes (*Ecclesiazousae*, 1002 ff.): 'Why should we buy a meat hook (13) for our jars, when we can let down such an old hag as this to grapple the pots from our wells?' In many of the narrow-mouthed jars used for drawing water a hole was cut in the shoulder to facilitate filling.

In addition to wells, many cisterns have been found around the Agora. Most date from the 4th to 1st centuries B.C. Aristotle's advice (*Politics*, III, 1330b, 3 ff.) 'that there should be (in the event of war) a good natural supply of water, but if not, it can be achieved by fashioning large and capacious cisterns for rainwater' is perhaps not the cause of the innovation but reflects the same climate of opinion. Cisterns are usually small, averaging about three meters in diameter at floor level, probably because of the faulty nature of the rock formation which also made necessary the flask-like shape, to minimize the risk of collapse. Occasionally two independent cisterns are joined by a channel (14), in the middle of which is a low barrier; when the water sank to a hardship level, each had only his own chamber to draw from. The inside walls of chambers and passages alike were covered with a fine hydraulic cement, often with toe-hole steps for ease of access. Vessels were often either dropped or even stored in cisterns when there was room (15). Perhaps it was for the retrieval of these that there existed the skill in diving and swimming in cisterns that Plato uses (*Laches*, 193C; *Protagoras*, 349E) as an example of the need for practice and preparation in any activity. We assume, at any rate, that the word *phrear* which he uses means cisterns in this context, since there

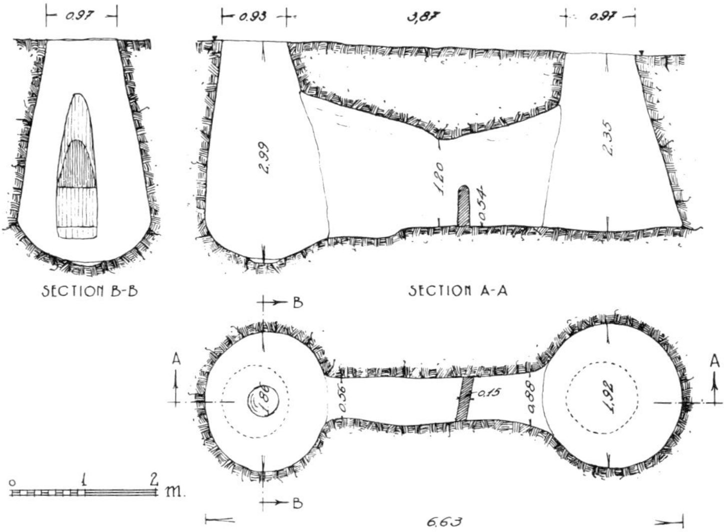


14. Cistern on Kolonos Agoraios. A bronze shield, booty from the Battle of Pylos 425 B.C., was found at the bottom; it had presumably served as a lid.

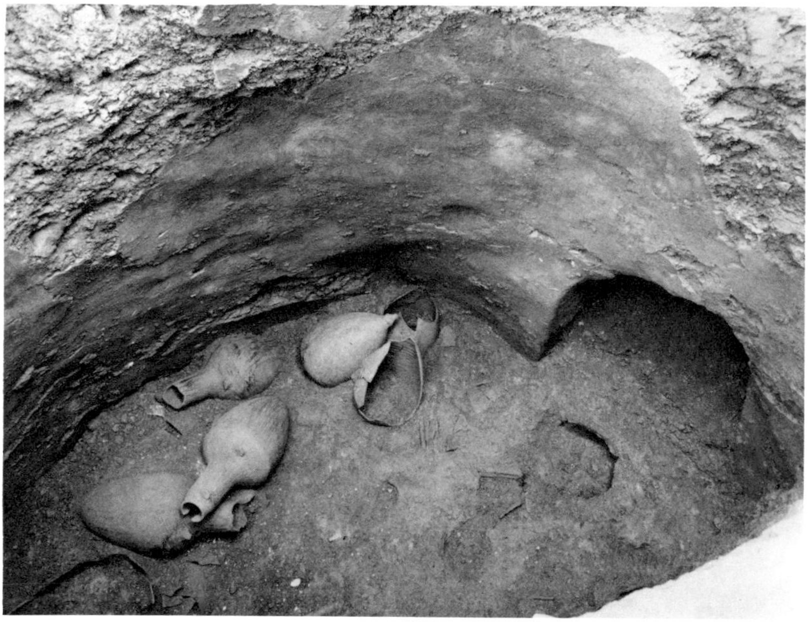

15. Cistern with Storage Amphoras. The passage leads to another chamber.

is so little scope for swimming in wells. But when Thucydides (II, 48) uses the same word for the water sources of the Peiraeus which were thought to have been poisoned by the enemy to cause the Great Plague, it is more likely, in the comparative absence of 5th century B.C. cisterns, to mean wells.

The water which was stored in cisterns would ordinarily have been rain-water, collected from the surrounding roofs which were drained into the eavestroughs and thence by pipe to channels at ground level flowing toward the cistern mouth. The cistern water, we may assume, was normally used for washing, the well water for drinking.

Wells and cisterns must most often represent private enterprise; bringing water from springs at a distance was necessarily a public-works project, whether instituted by Peisistratos in tyrannic beneficence or by a duly-elected magistrate. The earliest pipeline found in the Agora was that which lay beneath the main east-west road just to the south of the square; it belongs to the latter part of the 6th century B.C. (16). A narrow trench was first dug in the earth or soft bedrock, and the pipes were laid from east to west, showing that the water was being brought from the east. The direction is evident partly from the numbering of the pipes and partly from the consistent location of each pipe-section's lidded hole toward the east. That is, as each section of pipe was fitted by groove and flange (18) to the previous, the workman could reach in with one hand to seal the joint (16). For the structure of the pipes we should compare Vitruvius (VIII, 6, 8): 'Earthenware pipes are to be

16. Feed-pipe of the Enneakrounos.

made not less than two inches thick, and so tongued that they may enter into and fit one another.'

Two different kinds of pipe are used in this line. One series is carefully made, of buff clay, with roughly oval hand-holes and apparently numbered with painted inscriptions, perhaps to insure matching of joints, perhaps to measure distance, since from joint to joint the sections are very close to two (Greek) feet each. One section, for instance, has the fifth letter (epsilon) of the alphabet at its eastern end and the sixth (digamma) at its western; another pipe has the sixth letter at its eastern end and the seventh (zeta) at its western. All of these letters are accompanied by the same arbitrary squiggle which must distinguish this alphabetic series from others. But between the two sections which share the sixth letter another section intervenes with the eighteenth and nineteenth (koppa and rho) letters of the alphabet. This makes it likely either that the original workmen happily ignored the manufacturer's instructions or that they were re-using sections from another line which were not being delivered in strict numerical order. (That the confusion is not a result of repairs is witnessed by the complete absence of later disturbance in the stratification.) The admixture in the same pipeline of the second kind of pipe sections suggests the same two alternatives. The second type is less carefully made. The pipes are of gray clay, with square hand-holes and no numbers, but with what is probably the name of the maker (Charon) incised when the clay was still soft (17).

17. Feed-pipe of the Enneakrounos inscribed Charon.

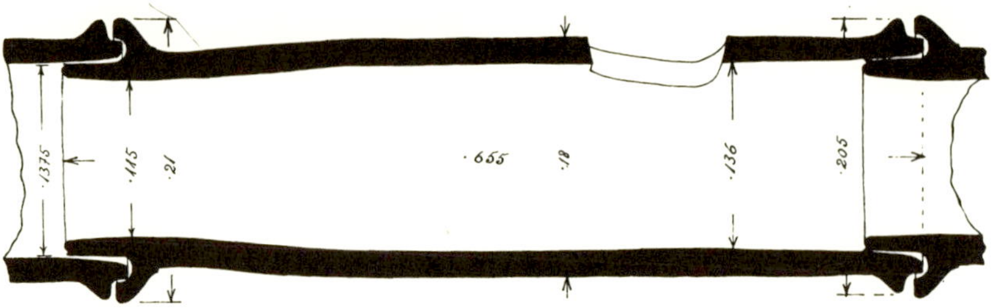


18. Waste-pipe of the Enneakrounos.

19. Stone Aqueduct of 4th Century B.C. A Bedrock; B Contemporary Road Surface; C Pipes of Turkish period.

20. Stone Aqueduct with Modifications.

A Original Water Channel
B Round Terracotta Pipe
C Ledge for Similar Pipe
D Square Terracotta Channel

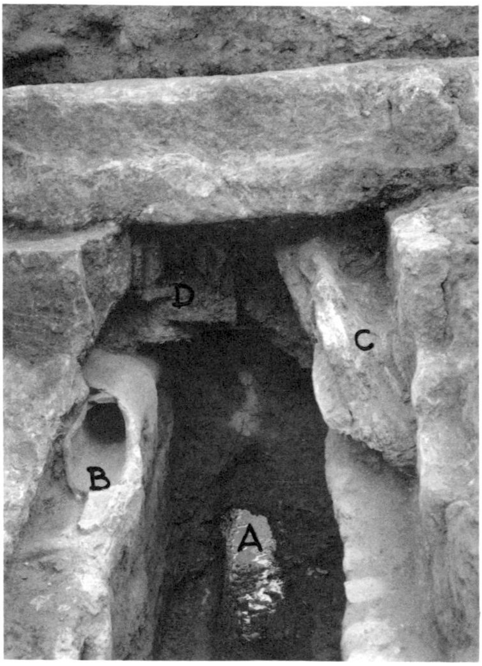


Inside the pipes is a hard lime deposit up to half an inch thick along the bottom and part way up the walls. It is clear from its date and position that the line was bringing water to the Southeast Fountain House, the Enneakrounos (2), even though the line is broken off by later disturbance and all trace of the intake is destroyed. The height of the line above the basins in the fountain house is such (about two meters) that it is reasonable to restore spouts set fairly high in the wall.

Succeeding the 6th century B.C. terracotta pipeline under the east-west road was a large stone aqueduct (19) which may be dated, partly by context pottery and partly by the increased water needs consequent on the building of the Southwest Fountain House, to the 4th century B.C. Made of large blocks of soft limestone set on end to form the walls and laid out crosswise for both floor and cover slabs, it is an impressive structure. The floor blocks are much thicker than the covers and have a channel some 0.20 m. deep cut in the center, leaving ledges on either side for passage by hand or foot. In later times (20) provision was made for three terracotta channels both to supplement the flow and also perhaps to bring in the water at a greater height to facilitate distribution from this gravity system.

The stone cover slabs are only a few centimeters below the 4th century B.C. surface of the east-west road. The aqueduct ran below the middle of the road, while on the surface along the southern edge ran an open stone gutter.

In addition to stone and terracotta, lead (21) was also used for the piping of water, although the comparative scarcity of the remains suggests that many of the ancient Athenians would have agreed with Vitruvius (VIII, 6, 10): 'Water-supply by earthenware pipes has these advantages. First, if any fault occurs in the work, anybody can repair it. Again, water is much more wholesome from earthenware pipes than from lead pipes. For it seems to be made injurious by lead, because white lead is produced by it; and this is said to be harmful to the human body.'

In addition to the more frequent gravity systems, pressure lines were sometimes used for column fountains and wherever it was necessary to make water run uphill. Ordinarily these lines employed pipes with thick walls and cemented joints but no hand-holes. The exceptional examples in 24 both come from the line (B) in 1. The section at right, opened up, may be seen to have an inner pipe of lead but no trace of leading at the joint. The section to the left had no inner pipe preserved but was jointed with a band of cement on top of which a band of lead had been poured.

Remnants of fountains to which water was brought by these and other pipelines are found with fair frequency in the Agora, but often in so ruinous and fragmentary a state that reconstruction must be, in part, conjectural. Of the minor fountains during the Greek period we may note the following: basin types in the square to the south of the Bouleuterion, at the south end of the Stoa of Attalos terrace, and in the niche in the back wall of South Stoa II (23); pressure-line fountains to the south of the Tholos and in front of the Propylon to the Bouleuterion. The great fountain houses of this time were

21. Lead Pipe.
2nd Century B.C.

22. Lead Screen from a Water-line. 2nd Century B.C.

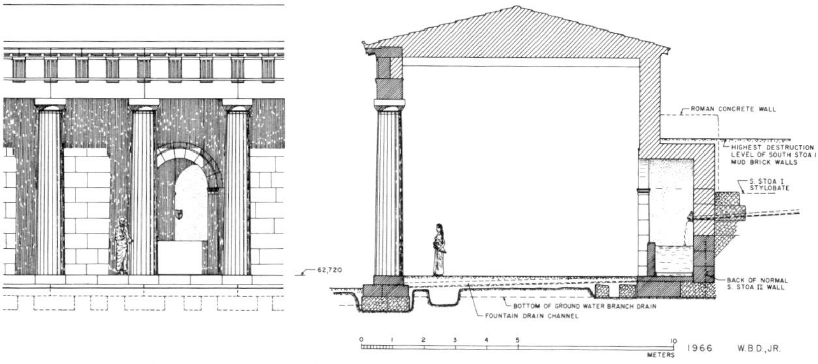


23. Fountain in South Stoa II. 2nd Century B.C.

at the southwest and southeast corners of the Agora. There is little or no certainty about their identification with named fountains of Athens, but Pausanias, at least, in his 2nd century A.D. guidebook, gave a name to the one at the southeast (I, 14, 1): 'Near by (the Odeion of Agrippa) is a fountain called Enneakrounos, since it was thus fitted out (with nine spouts) by Peisistratos.' The 6th century B.C. date of the original building provides confirmation of Pausanias' designation, as does his next point of reference: 'Above the fountain are temples, one of Demeter and Kore, the other of Triptolemos.' These unquestionably comprise the Eleusinion, above and beyond the South-

24. Sections of a pressure Pipeline: Terracotta and Lead.

25. Model of Southwest Fountain House. 4th Century B.C.

east Fountain House on the other side of the Panathenaic Way. But the waters of Enneakrounos are troubled not only by contradictory topographical evidence in other ancient authors but also by Pausanias himself when he says in this same connection: 'There are wells throughout the whole city, but this is the only spring.' The only evidence here of the spring Kallirrhoe, which authors from Thucydides on make the predecessor of Enneakrounos, is the unusual dampness of the area which necessitated special drainage in both ancient and modern times.

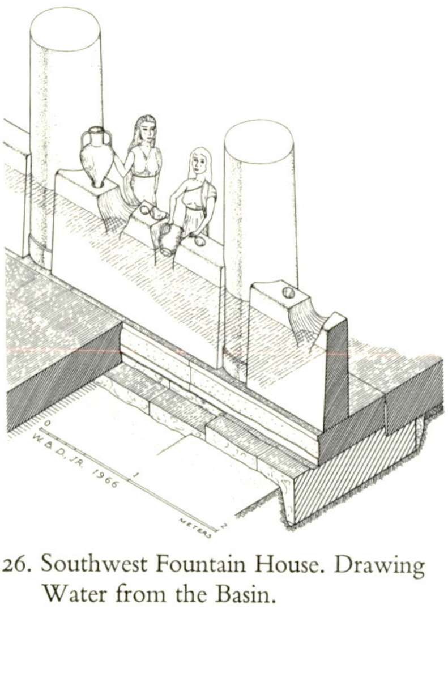

26. Southwest Fountain House. Drawing Water from the Basin.

The L-shaped plan of the Southwest Fountain House (25) is well adapted both to its position at a busy intersection and for the economical housing of extensive draw-basins (26). Additions to the original building provided spouts to supplement the basins.

Other fountains were built in the Agora during the Roman period when again the benevolence of despots could be exercised in popular public works. Small fountains accompanied both the Odeion of

27. Water Pitcher
with a Fountain Scene.
Late 6th Century B.C.

Agrippa and the Library of Pantainos. The aqueduct of Hadrian, finished in the reign of his successor and part of the Athenian water-system today, seems to have given a new spurt to fountain building in the Agora in the middle of the 2nd century A.D. The new abundance of water made possible its lavish and ornamental use in the Circular Fountain (28), the lead pipeline of which suggests the probability of a vertical jet, and in the Nymphaeum, which had a semicircular back wall with niches for sculpture (3).

28. Northeast Corner of the Agora. Model. The Round Fountain stands in front of the Stoa of Attalos.

29. Terracotta Waste-pipes of the Enneakrounos.

The overflow from the fountains was dealt with in various ways. The waste conduit from the Southeast Fountain House shows two periods: a smaller, earlier line (at left in 29) and a larger later line. The earlier line is carefully jointed with lids in place and seems to belong to the earliest period of the fountain. At that time apparently the overflow water was not just disposed of but kept clean for some secondary purpose. The situation was different when the original line was given up in favor of larger pipes. These latter look just like the original feed-line, and it may be that the new stone aqueduct not only brought in more water which required larger overflow pipes but also released the old feed-line pipes to be thus used. Furthermore, an increased supply of water in the Agora both here and in the Southwest Fountain House may have obviated the need for the overflow; this would also explain the carelessness of the way in which the later drain line was laid.

In both periods the pipes from east and west basins came together in a Y-pipe especially designed for this angle and position (30, 31): from the junction a single line continued north. The absence of lidded holes in this pipe section makes it certain that the primary use of the openings was in sealing joints, and not for cleaning out the pipes. Had the holes been for cleaning, one would be needed more in this double-elbow shape than in any straight pipe, but none is necessary here for sealing joints since it is possible easily to reach each of the arm ends from the stem (31).

30. Junction of Waste-pipes of the Enneakrounos. Later System. View from Southwest.

31. Y-pipe from the Waste-pipe of the Enneakrounos.

32. Water Channel bordering the Panathenaic Way. 2nd Century B.C.

From other fountains the water was carried off in open stone gutters with settling-basins at intervals (32) from which water could be dipped up for washing, scrubbing and watering of both animals and vegetation. So Plato recommended (*Laws*, VI, 761B–C; *Kritias*, 117) that the overflow of fountains be used to water the groves of the gods. In the Agora one of the stone channels with basins branches out toward the Altar of the Twelve Gods, around which planting holes give evidence of the trees that must once have flourished there.

In addition to the overflow from fountains that might be used for other purposes, there was of course much waste water that had to be drained off. Multitudes of small drains, using round terracotta pipes or open terracotta channels square in section, run from every building into larger drains under streets. But the drain most impressive both for size and age in the Agora is the so-called Great Drain (33), which ran in the beginning from the area of the Tholos north through the Agora, and beyond to the Eridanos. The very fine masonry of its walls combines with ceramic evidence to date its construction to the early 5th century B.C. That so capacious a drain should have been built in so early a period and in an area that only later became comparatively well supplied with water becomes understandable when one sees the way in which a heavy rain still turns the Great Drain into a rushing torrent. The channel is

33. The Great Drain of the Agora. Width of Channel one meter.

34. Wall of the Great Drain, West Branch.

35. Great Drain, West Branch, and Overpass.

about one meter in width and one in height. The existing cover slabs are much later than the original construction.

Extensions of the Drain to the southwest (34) and southeast became necessary about the turn from the 5th to the 4th century B.C. To the southwest even before the built drain there was a channel cut in the bedrock along the natural line of drainage. When it came to building, the situation was different from that in the public square; here the drain ran between private houses and had often to skirt the angles of private property. Also, the complete lack of uniformity in the walls of the drain from one house-property to another suggests that each householder was responsible for the construction of the drain along his property.

Where the southwest extension of the Great Drain passed under the main road leading to the Peiraeus gate (to the southwest of the Tholos), it was necessary to construct an overpass (35). By means of corbelling, the walls of the drain were brought within about 0.80 m. of each other (from a ground width of almost a meter and a half), so as to facilitate bridging. Drains from under the crossing road were also found to empty into the Drain here. In early Roman times elliptical tiles were laid in the channel to carry the water; a filling of gravel above the pipes allowed the upper part of the broad channel to be used as a road.

With the building of the Middle Stoa in the 2nd century B.C. the course of the southeast branch of the Great Drain had to be shifted so as to pass through the foundations the shortest way possible, or transversely (36). In the new course, the walls are of large stone slabs set upright; in places they rest on transverse floor slabs, elsewhere on the dressed bedrock. At one point a manhole is preserved (37), measuring 0.70 m. (the width of the drain itself) by 0.60 m.; the mortar with which the small stones forming its walls are held in place belongs to the Roman period.

36. Great Drain, East Branch: Passage through Foundations of Middle Stoa.

37. Great Drain, East Branch with Manhole.

38. Terracotta Drains of the Hellenistic Period South of the Middle Stoa.

When the Great Drain and its branches were no longer kept clear and in repair, the soil washed down from the heights to the south was spread broadcast over the whole area, the deposit being thickest along the natural lines of drainage to the southwest and the southeast. The effective simplicity of the original drainage system has been shown again since the excavation and clearing of the channels have restored them to working order.

Lesser drains which interrupt, cross, supplement and replace one another are found throughout the Agora. Some were open gutters beside the road, but more often they were covered and ran underneath with manholes for ease of access (38, 39). Continuous lines do not often survive in an area like that of the Agora where each succeeding century disturbs the remains of its predecessors.

39. Built Drain of Late Roman Period with Manhole.

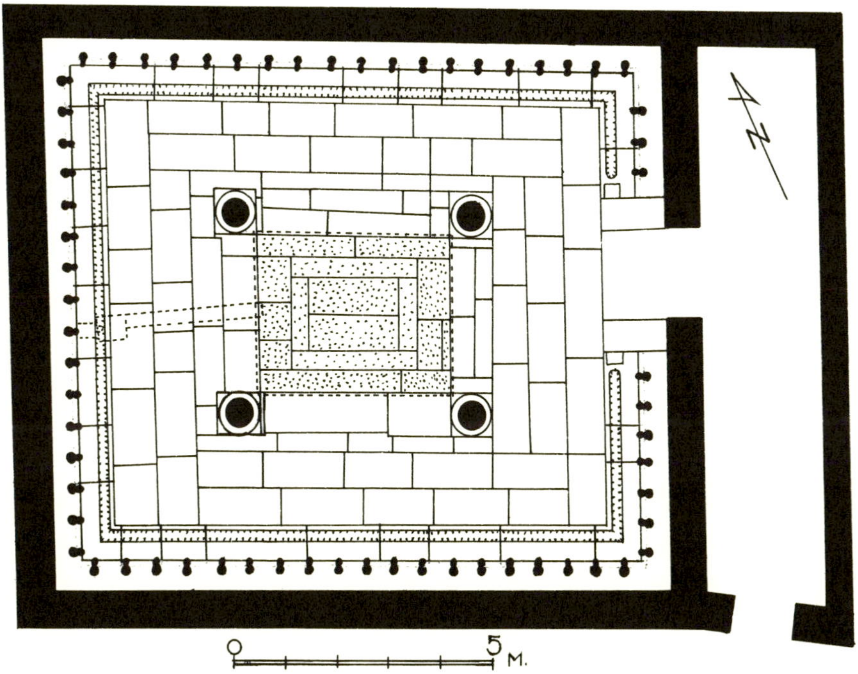


40. Plan of Latrine at the Entrance to the Roman Agora.

One essential of civilized urban life is intimately connected with drains: the public latrine. Fortunately the archaeological remains speak more clearly on this subject than do our literary sources. The two examples in the Greek Agora at the southwest and southeast entrances are far less well preserved than the fine example at the east entrance of the Roman Agora (40). The construction is simple, with a deep channel running around the four sides of a large room, the central part of which was open to the sky; over the channel were set marble slabs pierced with holes; the rather close-quarter seating capacity was 65, and the privacy non-existent (41), circumstances which give point to a fragment from Antiphanes, a comic poet of the 4th century B.C.: 'whoever thinks he's more than human, going to the public latrine, will see himself just like everyone else.' Private latrines were found in all the larger houses of the Roman period around the Agora, usually near the street door of the house.

Private bathing establishments, like domestic latrines, cannot be expected in the Agora, but there is at least some trace of a public bath, even though the circular foundations have not yet been found which would be suitable for

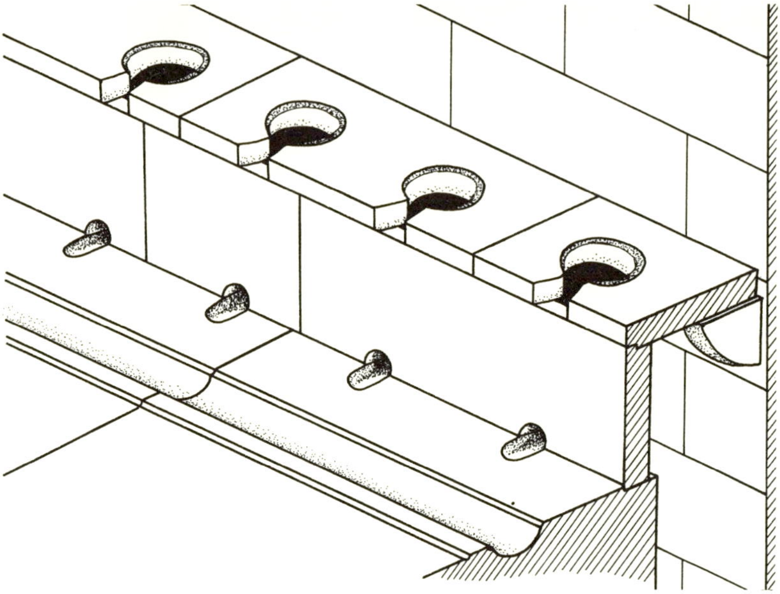

41. Detail of Latrine.

housing a series of tubs like that on the back cover of this booklet. This was found built into a 19th century house-foundation and is not likely to have been transported far for such a purpose. The curves of its short ends require a large circular room around the perimeter of which tubs were radially arranged with the heads to the outside and feet toward the inside. Such was the basic plan of baths of the classical period. Tubs were sometimes hollowed out of the rock, sometimes made of terracotta, sometimes built of small stones or bricks with a hydraulic cement coating. Other facilities often but not invariably available in Greek baths include deep pools for total immersion, wash basins and foot-baths. Water was provided by means of reservoirs, fountain spouts and sometimes wells. To judge from frequent remarks in the comic poets, hot baths were common even in the 5th century B.C.; for example, Hermippos (fr. 76 E): 'No, by Zeus, nor is it right for the true man to get drunk or take hot baths the way you do.' We know little about the arrangements for heating, partly because these were not part of the basic structure of the building as in Roman baths. But the 4th century B.C. comic poet Alexis (fr. 101 E) does touch on the horrid contingency 'if there was not fire on the hearths at the bath and if the oiling-room was locked.'

Among several baths of the Roman period that have come to light may be mentioned one in the area to the southwest of the Agora (42). The remains owe their preservation to the rubble and concrete construction required by the presence of heat and water in large quantities. For reasons of both convenience and economy, one supposes, the bath is built over the southwest branch of the Great Drain. It is made up of a courtyard, an undressing room (A for *apodyterion*), small service chambers in one of which was a well, then the warm room with arrangements for heating below and the circulation of hot air under the floor and inside the walls (T for *tepidarium*), and finally the hot room (C for *caldarium*) with hot plunge baths, the restoration of which is in this case conjectural but required by the comparative uniformity of hundreds of more complete examples. Among Vitruvius' remarks on the building of baths, we may note particularly (v, 11, 4): 'The size of the baths must depend upon the number of the population. Let the rooms be proportioned as follows, so that their width be one-third of their length, not counting the niches for wash basin and tub. The wash basin should always be placed under a window, so that those who stand around it do not obscure the light with their shadows.'

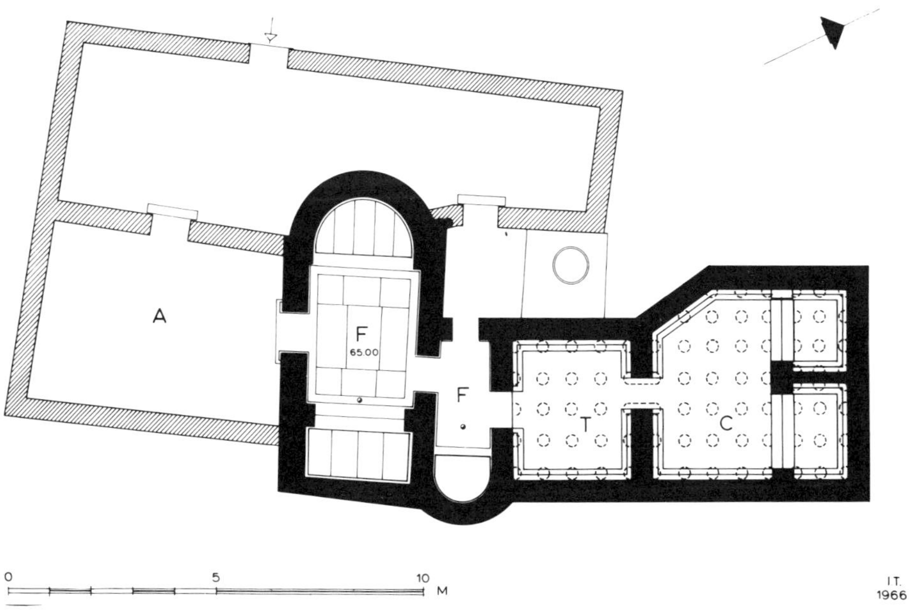


42. Plan of Roman Bath to the Southwest of the Agora.

43. Water Clock near the Southwest Entrance of the Agora.

Still another use for water in the Athenian Agora is the intriguing but not completely understood water clock built about the middle of the 4th century B.C. (43, 44). A shaft (A) lined with hydraulic cement must have been filled from a similarly coated reservoir (B). Much worn steps (C) leading down to the lower level give access to the metal-lined aperture (D), through which, when unplugged, the water would be drained from the shaft at a regular rate. The diameter of the metal tube and the dimensions of the shaft must have been calculated so that the sinking water-level accurately reflected the passage of the hours. The walls of the shaft itself may have been marked with the hours, but it is more likely, for visibility at a distance, that some kind of flag mounted on a float would be viewed against a calibrated background. Since the standard twelve hours between sunrise and sunset necessarily changed their length from season to season (ranging from perhaps 45 minutes at the

44. Detail of the Water Clock showing the bottom of the Shaft and the Water Outlet.

winter solstice to 72 at midsummer), frequent changes must have been made to adjust the rate of flow. Vitruvius (IX, 8, 6 ff.) describes mechanisms designed for this purpose.

Before the 4th century B.C., the times of day were commonly indicated in more general terms. Hear Xenophon describing Socrates (*Memorabilia* I, i, 10) 'But that one was always in the public eye; for early in the morning he frequented the walks and the gymnasia, and at the time when the market was filling he was always visible, and for the rest of the day he was wherever he would be with the most people.'

This rapid survey of the use and disposal of water in the Agora may fittingly end with a brief reference to the sacred properties of water and its symbolic use for purification. It is perhaps not surprising for so political a people as the Athenians that the civic center or gathering-place (Agora) should have had a

45. Marble Perirrhanterion.

sanctity as jealously guarded as that of the shrines of the gods. Perirrhanteria (basins for sprinkling water, 45) were set up at the entrances of the Agora as at the borders of sacred precincts. Aischines, an orator of the 4th century B.C., reminds the jury (III, 176): 'The lawgiver excludes the coward and the deserter from the perirrhanteria of the Agora.' Compare Hippokrates, Father of Medicine (*Malum sacrum*, IV): 'We ourselves set boundaries for the gods of their sacred places, so that no one may enter if he is not pure, and going in we sprinkle ourselves.'

46. Exhibition of Waterworks in the Stoa of Attalos.